T0025291

ON THE LINE

The Story of the Greenwich Meridian

ROYAL
MUSEUMS
GREENWICH

At the heart of the UNESCO World Heritage Site of Maritime Greenwich
are the four world-class attractions of Royal Museums Greenwich –
the National Maritime Museum, *Cutty Sark*, the Royal Observatory,
and the Queen's House.

The Royal Observatory Greenwich is the home of space and time:
the Prime Meridian and GMT, awe-inspiring astronomy and the
Peter Harrison Planetarium.

First published in 2003. This revised edition published in 2019 by the
National Maritime Museum, Park Row, Greenwich, London SE10 9NF.

9781906367619
Text © National Maritime Museum, London
Original text by Graham Dolan. Revised edition text by Louise Devoy
All photography © National Maritime Museum
with the exception of the following pages:
34 and 45 – Cambridge University Library
36, 37, 38, 40, 46–47 © Alamy
53 © NERC Space Geodesy Facility
Graphics and design by Matt Windsor

A CIP catalogue record for this book is available from the British Library.

Printed and bound in the UK by Gomer Press
10 9 8 7 6 5 4 3 2 1

Contents

Key terms

Astronomer Royal – the official astronomer of the Royal Observatory, Greenwich, who ran the site and was required to live in Flamsteed House. Today it is a purely ceremonial title awarded to a prestigious British astronomer with no connection to Greenwich.

ecliptic – the Earth's orbital plane around the Sun projected into space (onto the celestial sphere).

Greenwich Mean Time (GMT) – a time standard originally based on observations of the Sun as seen at Greenwich. Natural variations in solar time were smoothed out to create a regular, average (mean) length of time that could be used to set clocks.

latitude – the measure of how far north or south somewhere is from the equator from 0° to 90° at the Poles.

local meridian – an imaginary line connecting the North and South Poles, passing through the zenith at your location.

local solar noon – the moment when the Sun crosses your local meridian (north-south line).

longitude – the measure of how far east or west somewhere is from the Prime Meridian.

lunar distance method – a navigational technique whereby mariners measure the angle ('lunar distance') between the outer edge (limb) of the Moon and a bright star, noting the time. After several calculations, mariners can compare this local observation with the predicted timing of the same observation made elsewhere as 1 hour's time difference is equivalent to 15° longitude.

marine chronometer – timekeepers designed for navigation at sea.

marine sextant – a device used to determine longitude at sea, by measuring the angle between the edge of the Moon and a bright star, along with both their altitudes (heights) above the horizon.

Prime Meridian – a meridian that is chosen to be 0° longitude for common reference. The Greenwich Meridian was adopted as Prime Meridian in 1884.

Pole Star (Polaris) – the brightest star in the constellation of Ursa Minor and the closest star to the celestial North Pole. Its altitude (height) above the horizon is equivalent to your latitude, as measured with a sextant.

sidereal day – the period of the Earth's rotation measured from the vernal equinox crossing the local meridian. A mean sidereal day is 23 hours, 56 minutes and 4.09 seconds long.

solar day – the period of the Earth's rotation measured from the Sun crossing the local meridian. A mean solar day is 24 hours long.

UTC – Coordinated Universal Time (known as UTC) is the standard time by which the world regulates clocks and time.

vernal equinox – a point on the celestial sphere used for measuring sidereal (star) time. The Sun crosses this point between the celestial equator and the ecliptic around 21 March each year.

zenith – the highest point immediately above an observer, with an altitude of 90° above the horizon.

The first mention of each term in the text can be found in **bold**.

Introduction: Where east meets west

The Royal Observatory, Greenwich, is famous around the world as home to both **Greenwich Mean Time** (GMT) and the historic **Prime Meridian** (0° **longitude**). Just as the equator separates the northern and southern hemispheres, the Prime Meridian separates the eastern and western hemispheres. But what is a meridian? Why is the meridian at Greenwich so important? Why does it pass through an observatory and not a centre of government? Did the meridian exist before the Royal Observatory? And what is the link between the meridian and time? The answers lie in the history of navigation and the part played in this by the **Astronomers Royal** and others at Greenwich.

THE TRANSIT-CIRCLE AT THE ROYAL OBSERVATORY, GREENWICH.

▲ The historic Prime Meridian is defined as an imaginary line that runs from the North Pole to the South Pole, passing through the eyepiece of the Airy Transit Circle, a telescope situated within the Royal Observatory.

◄ Situated about 9 km east of central London, the Observatory attracts tourists from around the world who want to see the famous line that features on their maps and globes.

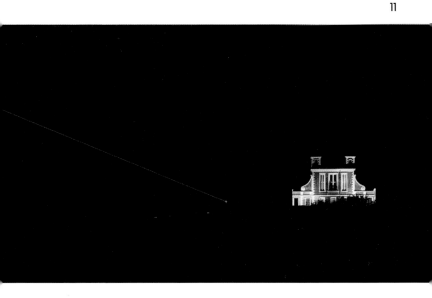

▲ The Prime Meridian is marked at night by a laser beam. The beam originates from a point above the Airy Transit Circle. Depending on atmospheric conditions, it can be seen from more than 20 km away.

◄ Although not marked in the Observatory courtyard until it became a museum, the Prime Meridian had been marked on the path outside much earlier. These visitors to Greenwich Park in the 1920s are about to cross the Meridian from east to west.

DID YOU KNOW?
The Royal Observatory was built on the foundations of an old castle, using mainly recycled materials from Tilbury Fort and the Tower of London. It cost just over £520, and was paid for by selling off old gunpowder.

What is latitude and longitude?

In order to pinpoint a particular location on a map of the world, you need to work out the coordinates of that location. Lines of **latitude** and longitude form the grid system used on globes, maps and charts. Latitude is a measure of how far north or south somewhere is from the equator, and longitude is a measure of how far east or west it is from a particular meridian. Although the equator is an obvious starting point from which to measure latitude, there is no natural equivalent from which to measure longitude. The north-south line passing through any particular point on the Earth's surface is known as the '**local meridian**'.

Lines of latitude all run parallel to the equator while lines of longitude (meridians) all converge at the Earth's North and South Poles. Latitude and longitude are both measured in degrees. Each degree of latitude corresponds to a distance on the Earth's surface of about 111 km. Each degree of longitude, however, corresponds to a distance that varies with latitude. The distance is about 111 km at the equator, reducing to 0 km at the poles.

For centuries, people have chosen different meridians to be their prime meridian (0° longitude). The Ancient Greeks chose the Canary Islands (known as the Fortunate Islands at that time) as theirs as it was the furthest landmass known to the west. Over the centuries, different cartographers adopted different locations for the prime meridian. In some instances, they chose islands further west across the Atlantic, such as the Azores or Cape Verde archipelago. Others chose places such as the south-west tip of Cornwall, known as the Lizard, or the meridian defined by astronomers at the Paris Observatory.

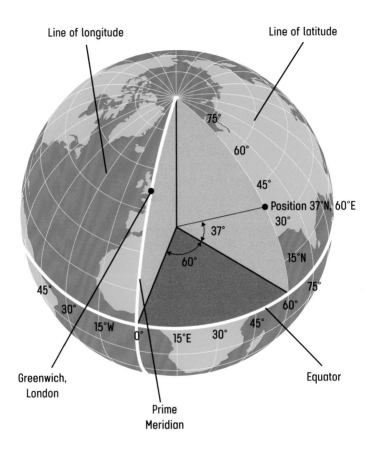

Line of longitude

Line of latitude

75°
60°
45°
● Position 37°N, 60°E
30°
37°
60°
15°N
45°
30°
75°
60°
15°W
45°
0°   15°E   30°

Greenwich,
London

Equator

Prime
Meridian

▲ Latitude and longitude are measured in degrees from the centre of the
Earth. In this diagram, the position of the marker on the map is 60° east of
the Greenwich Meridian, and 37° north of the equator.

Here are some examples of historic maps showing different meridians.

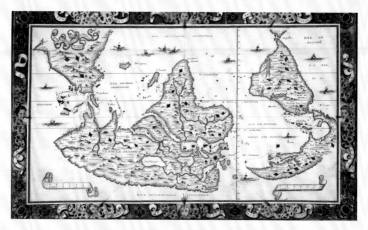

▲ Produced on vellum, this hand-drawn coloured map from 1567 was orientated with south at the top. The prime meridian extended through the Canarian island of El Hierro (known as Fero at this time).

▶ (Top) Produced by Willem Janszoon Blaeu in 1606, this world map shows the prime meridian passing through the Azores.

▶ (Bottom) In 1686 French cartographer Pierre du Val continued to use a prime meridian passing through the Canary Islands for this map.

DID YOU KNOW?
Longitude on Mars is measured from a prime meridian that passes through a small crater called Airy-0, named after the famous Greenwich astronomer.

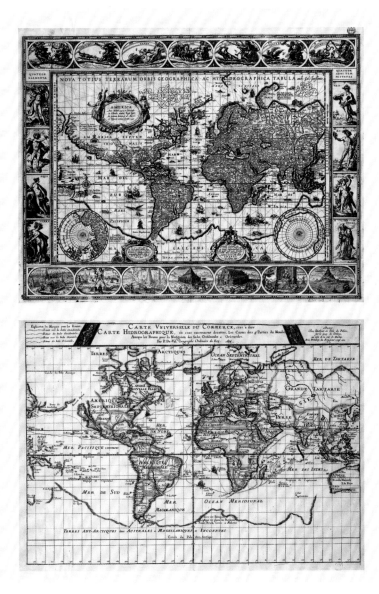

Celestial latitude and longitude

Astronomers use a similar grid system to plot the positions of the stars. If you imagine the Earth in space, we can project lines onto the sky that correspond to the equator, the North and South Poles and the apparent path of the Sun across the sky over the course of a year (**ecliptic**).

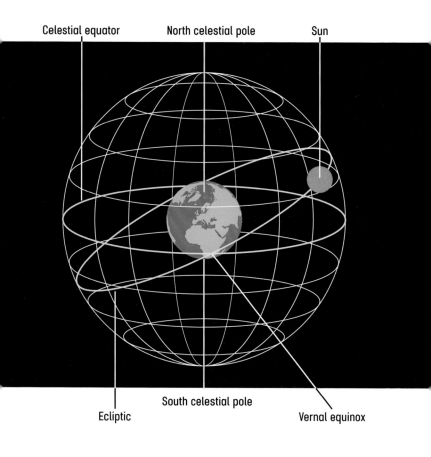

Converting the sphere into a flat rectangular map of the stars helps astronomers create a coordinate system. A star's north-south position, or celestial latitude, is called declination (Dec) and is measured in 0° to 90° above (+) or below (-) the celestial equator. A star's east-west position, or celestial longitude, is called right ascension (RA) and is measured in hours, minutes and seconds from 0 to 24 hours with subdivisions in minutes and seconds. The celestial equivalent of the Prime Meridian, 0 hrs RA, is the **vernal (spring) equinox**, historically known as the First Point of Aries. The Sun reaches this point around 21 March each year.

This coordinate system means that astronomers can map the positions of the stars and use the sky as a giant clock. As the Earth rotates on its axis once every 24 hours, the stars appear to move across the sky in a similar way to the Sun. They reach their highest point (culmination) when crossing the local meridian. Astronomers use a telescope to determine a star's declination by measuring its height above the horizon (altitude). They also use a clock to measure the star's right ascension. Once a star's position is known, it can be used to determine the exact local time on each occasion it re-crosses the meridian. Astronomers use this data to calibrate their clocks.

◀ If you imagine a celestial sphere around the globe, it will have lines similar to the lines of longitude and latitude on Earth.

▼ As the stars appear to move across the sky from east to west, astronomers use this coordinate system to time the precise moment when certain stars cross the local meridian.

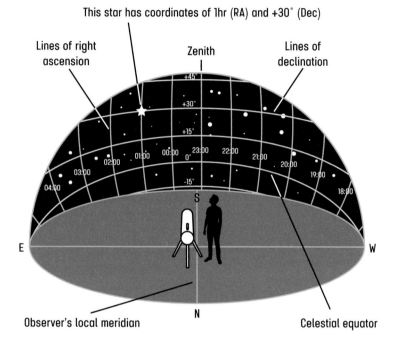

This star has coordinates of 1hr (RA) and +30° (Dec)

Lines of right ascension

Zenith

Lines of declination

+45°
+30°
+15°
0°
-15°

00:00 23:00 22:00 21:00 20:00 19:00 18:00
01:00 02:00 03:00 04:00

Z
S
E
W
N

Observer's local meridian

Celestial equator

But these are no ordinary clocks. While we generally rely on the Sun as our natural timekeeper, astronomers prefer to use sidereal (star) time instead as it is more consistent. As the Earth spins on its axis, astronomers can use a star's right ascension (RA) as a measure of sidereal time as it crosses the meridian. The **sidereal day** begins and ends when the vernal equinox crosses the meridian. Because of the difference between sidereal and solar time, each sidereal hour, minute and second is slightly shorter than its solar equivalent.

Sidereal day versus solar day

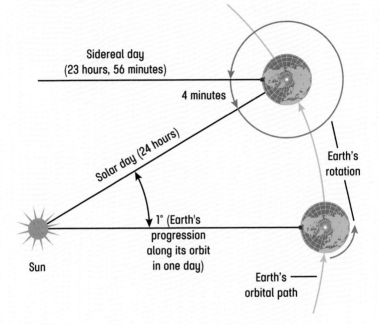

▲ There's a 4-minute difference between the sidereal and **solar day**. It takes a time interval of 23 hours, 56 minutes and 4.09 seconds for the vernal equinox (0 hrs RA) to cross the meridian (red dot) twice. But the time interval between two transits of the Sun (**local solar noon**) is 24 hours because the Earth is gradually moving around its orbit by 1° each day (360° in a complete year). The astronomer has to wait for the Earth to rotate for an extra 4 minutes for the Sun to cross the meridian again.

▶ Not all clocks are marked in hours. This clock was designed by John Flamsteed, first Astronomer Royal, to show the time in degrees, minutes and seconds of arc, according to the Earth's rotation relative to the stars.

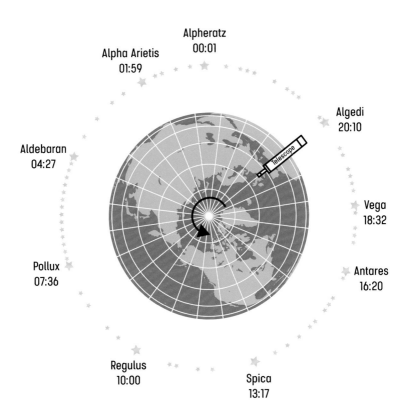

▲ To help improve their measure of sidereal time, astronomers at Greenwich used certain 'clock' stars whose positions were well-known. As the Earth spins on its axis, the telescope sweeps across the sky. The telescope was used like the hour hand of a clock, while a group of over 60 stars were used like the numbers around the dial.

Early meridians at Greenwich

Each astronomical instrument at the Royal Observatory was installed and aligned on its own local meridian (north-south line). Over three centuries of work at the Observatory, different Astronomers Royal installed newer, better instruments across the site to enhance their work. The difference in time and position between the old and new meridians was too small to be measured with the clocks and navigational instruments available at the time so there was little confusion.

Some of these instruments and their associated meridians were formally recognised as the 'Greenwich Meridian' many decades before the Airy Transit Circle was adopted as the world's prime meridian in 1884. For example, the first printed chart known to have used Greenwich as its prime meridian was published nearly 150 years earlier in *A Description of the Sea Coast of England and Wales* (1738) by Samuel Fearon and John Eyes. This chart was based on the meridian defined by the second Astronomer Royal, Edmond Halley (1656–1742), who used a 5-foot long transit telescope on the north-west edge of the Observatory from 1721 onwards. His successor, James Bradley (1693–1762), organised the construction of a more spacious building to accommodate an improved 8-foot long transit telescope that defined a new meridian from 1750 onwards.

▶   Halley's 5-foot (153 cm) telescope was used to define the Greenwich Meridian from 1721–1750.

▲ The Bradley meridian today with a 10-foot transit telescope made by Edward Troughton in 1816.

Over 15 years later, the fifth Astronomer Royal, Nevil Maskelyne (1732–1811) continued to use Bradley's meridian for defining the astronomical data that was later published as the *Nautical Almanac,* for the year 1767. Updated each year, this book of predicted observations became an essential tool for navigators at sea using the lunar distance method. In 1791, the Ordnance Survey (OS) used the Bradley meridian as the base line for their map of England and Wales; it still provides the basis of our OS maps today. The Bradley meridian was eventually replaced in 1851 with a new meridian defined by the Airy Transit Circle.

▲ The second Astronomer Royal, Edmond Halley.

▲ The third Astronomer Royal, James Bradley.

▲ *General Survey of England and Wales*, the first Ordnance Survey map, 1801, showing the section around Greenwich.

The longitude problem

To complete a safe and efficient passage at sea, it is essential to
accurately know your latitude and longitude. When Christopher
Columbus sailed across the Atlantic in 1492 there was no reliable
way of measuring a ship's position once out of sight of familiar
coastlines. Sailors could estimate their latitude by measuring the
angle of the Sun and stars above the horizon but there was no
equivalent measure of longitude. They had to rely on measuring
their speed and direction (bearing) to gauge their position,
a technique known as 'dead reckoning'. As more trade routes
opened up in the 17th century, this problem became more serious.
Journeys often took longer than expected and could end in
complete disaster if a ship got lost or ran aground. Some maritime
nations offered large rewards to anyone who could find a reliable
way of measuring longitude at sea.

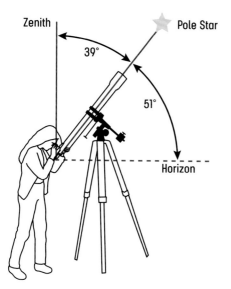

◀ An observer's latitude
on Earth is roughly
equal to the elevation
of the **Pole Star** above
their local horizon in the
northern hemisphere.
For example, the latitude
of London is 51.5°N so
the Pole Star is visible
51.5° above the horizon.

How does time help you measure your position?

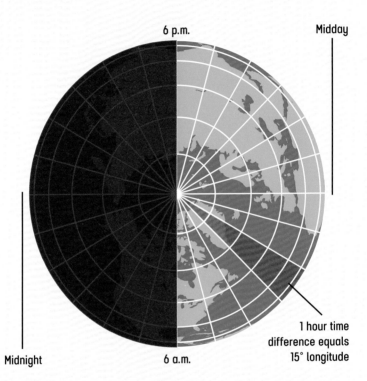

6 p.m.

Midday

Midnight

6 a.m.

1 hour time difference equals 15° longitude

▲ It all depends on the rotation of the Earth. Every 24 hours the Earth makes one complete turn on its axis (360°). This means that the Earth rotates 1° in four minutes and 15° in one hour. So to find out how far east or west a sailor was from home, they had to compare their local time with the time back home at exactly the same moment. Simple in theory but very difficult in practice!

A CHART
of the Southern Part of
SOUTH AMERICA
With the Track of the Centurion from the Island of St. Catherines to the Island of Joan Fernandes In which is inserted the Variation and Soundings observed on board her, together with her Deviation from her estimated Course in passing round Cape Horn, occasioned by the force of the Current.

◀▶ The British admiral George Anson was so uncertain of his longitude when rounding Cape Horn in the *Centurion* in 1741 that he ended up wasting a lot of time zigzagging back and forth in search of a safe passage. During this time, 70 sailors died of scurvy.

As long ago as 1514, it had been proposed that the Moon could be used as a clock for measuring longitude at sea. Over 150 years later, the Royal Observatory was built at Greenwich in 1675, in order to turn the so-called '**lunar distance method**' into a working solution – a task so complex that it took another century to complete.

The Moon's position against the background of stars changes in a complicated but predictable manner. In 1766, the fifth Astronomer Royal, Nevil Maskelyne, published the first *Nautical Almanac*. It contained a set of tables showing where an observer would see the Moon during 1767 if he could be positioned at the centre of the Earth. The Moon's angular position relative to nearby bright stars was listed at three-hourly intervals of Greenwich time.

To determine their longitude, a sailor had to measure the angle between the outer edge (limb) of the Moon and a listed star – the lunar distance – along with both their altitudes above the horizon. This was done using a **marine sextant**. Next, they had to do several calculations, particularly how this angle would change if observed from the centre of the Earth, as measured between the centres of the Sun and Moon. These corrections helped the sailor to compare his results with those predicted for an observer at the Royal Observatory, as listed for the same date in the *Nautical Almanac*.

From the time difference between these same observations, the sailor could calculate their difference in longitude from Greenwich. It was a challenging task that required careful measurement of angles in the sky – not easy on a moving ship followed by several hours of calculations.

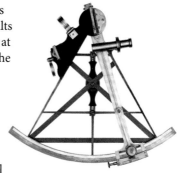

▲ An early marine sextant.

While the lunar distance method was still under development, the British Government, in 1714, set up the Board of Longitude. It offered a reward of up to £20,000 (equivalent to several million pounds today) to anyone who could find a means of measuring longitude at sea to the nearest half degree. It was eventually won by the British clockmaker John Harrison, who, at his fourth attempt, managed to build a timekeeper that would keep good time at sea. Now called H4, it was first tested at sea in 1761. Today, timekeepers designed for navigation at sea are called **marine chronometers**.

▶ The *Nautical Almanac* was developed from lunar tables published by the German mathematician Tobias Mayer, who relied on data provided by the third Astronomer Royal, James Bradley.

◀ Harrison's successful marine timekeeper, H4.

Distances of ☽'s Center from Stars, and from ☉ east of her.

Days.	Stars Names.	Noon.	3 Hours.	6 Hours.	9 Hours.
		° ′ ″	° ′ ″	° ′ ″	° ′ ″
1 2 3	α Pegasi.	46. 41. 15 33. 15. 35	44. 57. 51 31. 40. 16	43. 14. 53 30. 6. 42	41. 32. 32 28. 35. 2
4 5	α Arietis.	57. 55. 16 43. 32. 47	56. 6. 21 41. 46. 31	54. 17. 44 40. 0. 36	52. 29. 25 38. 15. 1
6 7 8 9	Aldebaran.	62. 4. 49 48. 36. 32 35. 37. 28 23. 22. 20	60. 22. 21 46. 57. 27 34. 2. 38 21. 55. 18	58. 40. 17 45. 18. 47 32. 28. 29 20. 30. 0	56. 58. 37 43. 40. 35 30. 55. 5 19. 7. 3
10 11	Pollux.	51. 3. 14 38. 27. 43	49. 27. 59 36. 54. 20	47. 52. 57 35. 21. 12	46. 18. 9 33. 48. 17
12 13 14 15	Regulus.	62. 42. 22 50. 23. 35 38. 13. 40 26. 11. 51	61. 9. 30 48. 51. 53 36. 43. 0 24. 42. 9	59. 36. 47 47. 20. 18 35. 12. 28 23. 12. 34	58. 4. 13 45. 48. 52 33. 42. 3 21. 43. 10
16 17 18 19 20	Spica ♍	68. 17. 41 56. 26. 28 44. 38. 16 32. 50. 51 21. 2. 16	66. 48. 34 54. 57. 51 43. 9. 50 31. 22. 21 19. 33. 33	65. 19. 30 53. 29. 15 41. 41. 25 29. 53. 51 18. 4. 47	63. 50. 31 52. 0. 41 40. 13. 0 28. 25. 19 16. 36. 0
21 22	Antares.	54. 40. 6 42. 27. 36	53. 9. 18 40. 54. 57	51. 38. 17 39. 22. 2	50. 7. 5 37. 48. 50
20 21 22 23 24 25 26	The Sun.	120. 36. 39 109. 36. 50 98. 25. 11 86. 56. 45 75. 6. 56 62. 51. 46 50. 8. 25	119. 14. 38 108. 13. 39 97. 0. 7 85. 29. 15 73. 36. 29 61. 17. 54 48. 30. 56	117. 52. 30 106. 50. 14 95. 34. 48 84. 1. 25 72. 5. 38 59. 43. 36 46. 53. 0	116. 30. 15 105. 26. 38 94. 9. 12 82. 33. 14 70. 34. 23 58. 8. 51 45. 14. 36

One line for the world

The 1770s marked a turning point in navigation. After years of uncertainty about their longitude at sea, sailors now had two ways of measuring it. Both worked by measuring time differences. The lunar distance method, using a marine sextant and the *Nautical Almanac*, always gave time differences from the Greenwich Meridian. Marine chronometers, on the other hand, could be set as the 'home' time so navigators could measure their position directly from Paris, Berlin, or any other city with a time standard.

For many years, different countries measured longitude from different meridians. The French and Algerians, for example, used the Paris Meridian, and the Swedes measured from one that passed through Stockholm. By the 1880s, many people could see the advantages of measuring from a single meridian. As a result, the International Meridian Conference took place in 1884 in Washington, D.C. After several weeks of discussion, delegates voted to recommend the Greenwich Meridian as the Prime Meridian of the world. Thanks to the continuing use of the *Nautical Almanac*, nearly two-thirds of the world's ships were already using charts based on the Greenwich Meridian, so it offered a pragmatic choice with minimal disruption.

Delegates at the International Meridian Conference made the following recommendation: 'the Conference proposes to the Governments here represented the adoption of the meridian passing through the centre of the transit instrument at the Observatory of Greenwich as the initial meridian for longitude.' This instrument was the Airy Transit Circle that had already been in use since January 1851 as the defining instrument of the Greenwich Meridian. Named after its designer, the seventh Astronomer Royal, George Biddell Airy (1801–1892), the instrument consisted of a 12-foot (3.7m) long telescope mounted

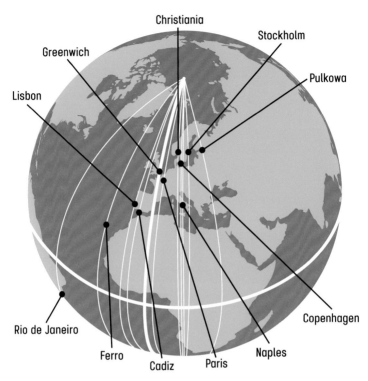

▲ In the early 1880s sailors measured longitude from different places. The most commonly used zero meridians are shown here.

DID YOU KNOW?
The length of the metre was defined in 1791 as 1/10,000,000 of the distance from the North Pole to the equator on the Paris Meridian. Since 1983 it has been defined as the distance travelled by light in a vacuum during a time interval of 1/299,792,458 of a second.

between two stone piers. Once the roof hatches had been opened, the astronomer sat in the pit below and looked through the eyepiece to time the precise moment at which certain stars appeared to transit the meridian. A few years later, Airy added electric time signals and a recording drum to help staff record their observations more effectively. Apart from a brief pause during the Second World War, the Airy Transit Circle was used for over a century to make around 600,000 observations.

▲ The seventh Astronomer Royal, George Biddell Airy, after whom the Airy Transit Circle is named.

▼ Delegates of the International Meridian Conference, Washington D.C., 1884.

▼ Sitting in the pit underneath the telescope, the astronomer timed the stars as they crossed the meridian.

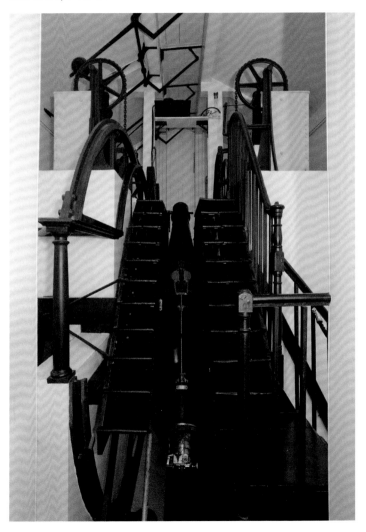

The Prime Meridian through the UK

The Prime Meridian passes through the centre of just a few towns in the UK. From coast to coast, it covers a distance of about 330 km. Its position is marked by a variety of monuments and signs. The most southerly point in the UK is Peacehaven in Sussex. Travelling north from the Observatory, the Meridian crosses the River Thames, passes up the Lee Valley and then runs close to the Roman road, Ermine Street, before entering the sparsely populated fenlands of Cambridgeshire and Lincolnshire. It continues across the Lincolnshire Wolds before crossing the Humber Estuary into Yorkshire.

▲ At the northern point, the meridian enters the coastal village of Tunstall, Yorkshire. This sign is located a few kilometres south, near Withernsea.

▶ Inhabitants of Peacehaven, East Sussex, erected an obelisk in 1936 to commemorate the reign of King George V and to mark the town's position on the Prime Meridian.

▲ Those approaching Greenwich through the Docklands can admire this meridian sculpture by David Dudgeon.

DID YOU KNOW?
Due to their convergence at the North Pole, the Airy and Bradley Meridians are about 30 cm closer where they leave the Yorkshire coast than they are at Greenwich.

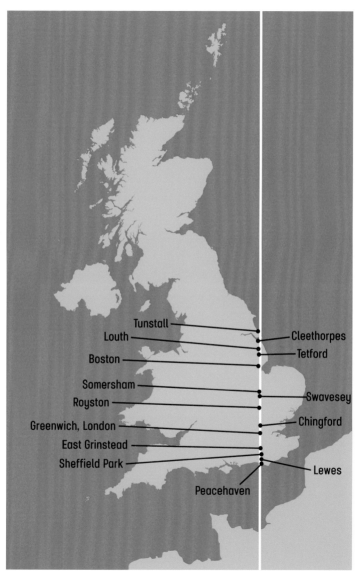

Tunstall
Louth
Boston
Somersham
Royston
Greenwich, London
East Grinstead
Sheffield Park
Peacehaven

Cleethorpes
Tetford
Swavesey
Chingford
Lewes

The Prime Meridian through the world

From pole to pole, the Prime Meridian covers a distance of 20,000 km. In the northern hemisphere, it passes through the UK, France and Spain in Europe, and Algeria, Mali, Burkina Faso, Togo and Ghana in Africa. The only landmass crossed by the Meridian in the southern hemisphere is Antarctica. Although the Prime Meridian passes through eight countries and crosses three continents, for nearly two thirds of its length it passes over the sea.

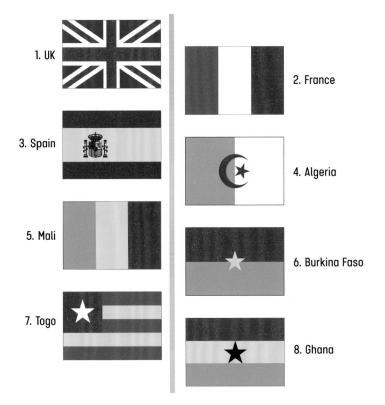

1. UK

2. France

3. Spain

4. Algeria

5. Mali

6. Burkina Faso

7. Togo

8. Ghana

The northern end of the Prime Meridian is the North Pole

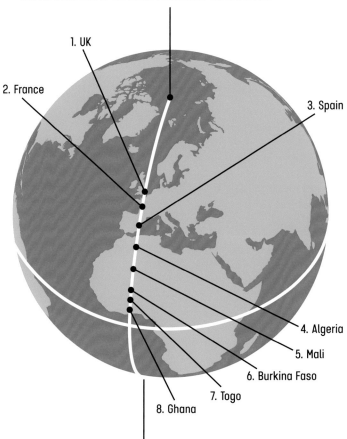

1. UK

2. France

3. Spain

4. Algeria

5. Mali

6. Burkina Faso

7. Togo

8. Ghana

At the very end of the line is the South Pole. No matter in which direction people standing at the South Pole look, they will always be facing north. Conversely, a person standing at the North Pole will always be facing south.

Greenwich Mean Time (GMT)

Delegates at the International Meridian Conference of 1884 also discussed the adoption of common time standards to help improve global communication and travel. Over the subsequent decades, governments across the globe gradually created time standards that were defined in relation to Greenwich Mean Time (GMT), similar to the use of the Prime Meridian for mapping.

But GMT was nothing new. It was originally created in the 1670s by the first Astronomer Royal, John Flamsteed (1646–1719), but was only used by other astronomers and navigators. The time when the Sun crosses over the local meridian is called the 'local solar noon' or 'local noon'. By using accurate pendulum clocks, Flamsteed realised that the length of each 'natural' day measured from one local noon to the next varies slightly through the year. This difference is due to the tilt of the Earth on its axis and its elliptical orbit around the Sun, rather than variations in the rate at which it is spinning. To create a regular 24-hour day that can be maintained by a clock throughout the year, Flamsteed formulated an 'equation of time' to create an average or 'mean' length of a natural day. When natural days are shorter than average, a clock will seem to run a few minutes slow compared to a sundial. When they are longer, it will appear to gain. The term 'Greenwich Mean Time (GMT)' describes the clock time based on the meridian measurements of the Sun and stars, as seen by an observer at Greenwich.

DID YOU KNOW?
We still use sundial terms in our daily language. The terms a.m. and p.m., referring to morning and afternoon, are abbreviations of ante meridiem (before the meridian) and post meridiem (after the meridian).

▼ Portable sundials have a compass so that they can be aligned along the local meridian before being read. This sundial can be used at different latitudes by adjusting the bird's beak on the gnomon (shadow caster) and using the relevant hour scale from 40°–52° latitude.

Most people relied on public clocks in town centres that were set to local mean time, based on local sundial time ('local apparent time') that had been adjusted by the equation of time. In Britain, this meant that local mean time varied with a time difference of nearly 30 minutes from Great Yarmouth in the east to Penzance in the west.

Variations in local time from east to west made little difference to people's lives until the development of the railway networks. With each town on the railway line following its own local time, the organisation of railway timetables became chaotic and potentially dangerous. During the mid-19th century, each of the different railway networks in the United States kept its own timetable based on the local time of the company's home station, creating over 60 different time standards. A traveller journeying from Maine to California would have to change their watch at least 20 times during the trip to ensure making a connection. In Britain, the time variation was less extreme, but still necessary. By the 1840s, most railway companies had adopted Greenwich Mean Time (GMT) as their standard time across the network, irrespective of local time.

DID YOU KNOW?
There's a clock in Bristol that still has two minute hands: a black one to show local time (about 10 minutes behind) and a red one to show Greenwich Mean Time. It was installed in 1822 and was later adapted for railway users.

▶ A map from the 1850s showing the different local times across the UK before standard time was introduced. Railway passengers travelling from London to Liverpool had to put their watches back 12 minutes on arrival to compensate from travelling from east to west.

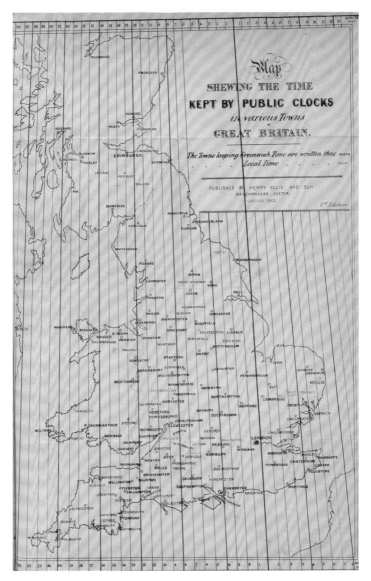

International time zones

The international time zone system is centred on the Prime Meridian, with the world's countries fitting loosely into a grid of 24 time zones, each 15° of longitude wide. The time in each zone is an hour ahead of the zone to its west and an hour behind the zone to its east. For example, when the time is 2.00 p.m. in Italy, it is 1.00 p.m. in the UK and 3.00 p.m. in Greece.

Most of the large countries that straddle several times zones, such as the USA, Australia and Russia, keep quite strictly to this

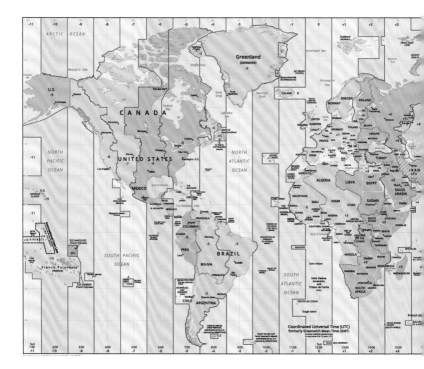

system; others such as China use one standard time across the whole country. Regional variations are usually decided by political or geographical factors. Occasionally, as in the case of India, time differences are based on half hours rather than whole hours. Some countries make use of Summer Time (also known as Daylight Saving Time). Clocks are put forward by an hour in the spring and back again in the autumn. About 60 countries use some form of daylight saving, particularly those furthest away from the equator where they have the greatest seasonal variation in daylight hours.

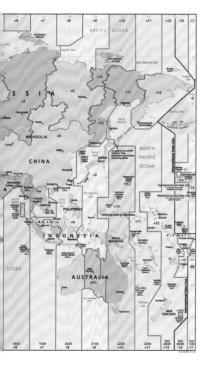

◄ The world's time zones. The International Date Line lies on the opposite side of the world to the Prime Meridian at longitude 180°. People crossing the line in a westerly direction skip forward a day. Those travelling in an easterly direction repeat the day.

DID YOU KNOW?
The islands of the Republic of Kiribati in the mid Pacific straddled the International Date Line until 1995 when its government decided to move the line eastwards so that Kiribati was no longer in two different days. It meant that Kiribati could trade more easily with Australia and New Zealand as they share the same day.

A new era of satellite technology

If you visit the Royal Observatory today, you'll notice that your smartphone does not show exactly 0° longitude when you stand on the historic Prime Meridian. This meridian has been superseded for most practical purposes by the International Reference Meridian (IRM). This meridian was created in 1984 using space-based satellite technology to calculate its position based on the Earth's centre of gravity. The position of the IRM was calculated precisely to ensure that the time measured from the historic Prime Meridian using astronomical methods was exactly the same as time measured from the new meridian using satellite technology. This means the position of the IRM is actually based on the position of the Greenwich meridian. It runs about 102 metres to the east of the Airy Transit Circle at Greenwich but varies at other latitudes.

Unlike the Airy Meridian, marking the position of the IRM on the ground is not feasible. The outer layer of the Earth is made of several tectonic plates which move in different directions very slowly. While the position of the IRM is fixed in space, the tectonic plates slowly move relative to the IRM. This means the satellite navigation coordinates of a specific location on the Earth's surface change over time, so other coordinate systems are used instead.

▶ Satellite navigation is so accurate that it has an average error of just a few metres – 3,000 times more accurate than the lunar distance method, which only allowed a sailor to measure his position to around the nearest 30 km.

DID YOU KNOW?
The Earth's tectonic plates are moving relative to one another at about the rate at which fingernails grow.

Timekeeping remains an essential part of navigation in our daily lives. The US Navy began to experiment with early satellite navigation systems known as TRANSIT during the early 1960s. They needed to accurately locate their submarines to ensure that the Polaris ballistic missiles were fired at the correct target, if necessary. The system was based on a network of five satellites orbiting 1,000 km above Earth every two hours. Submariners used the changing frequency of each satellite's radio signal as it crossed the sky as a means of measuring their position.

In 1973, the US Defense Department started to develop the Global Positioning System (GPS) and by 1994 the system was fully operational. A network of 24 solar-powered satellites, orbiting at an altitude of 20,350 km, transmit low power radio signals set by several atomic clocks on board the spacecraft. By timing how long it takes for the radio signals to arrive from several satellites in view, the receiver in your device can calculate your position to within a few metres. The more accurate the clocks, the more accurate the location for the user.

Today, other countries and international collaborations have their own alternative satellite navigation systems, either in use or in development, such as GLONASS (Russia), GALILEO (European Union) and BeiDOU (China). Despite our growing dependency on satellite navigation, many military services are still teaching traditional celestial navigation as a precaution against satellite failure caused by electronic warfare or collisions with debris from old satellites.

DID YOU KNOW?
*Satellite navigation is based on the simple equation **distance from satellite = speed x travel time of radio signal** (where the speed of the signal is equivalent to the speed of light at 299,792,458 metres per second).*

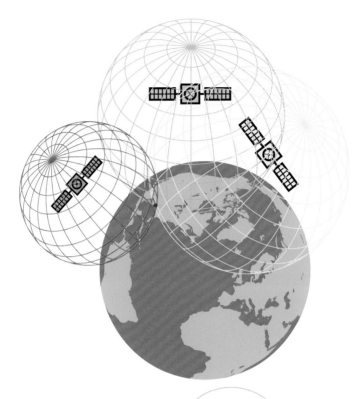

▶ The satellites broadcast
their signals in all directions.
By using at least 3 signals
('trilateration'), you can
identify your location as the
intersection point.

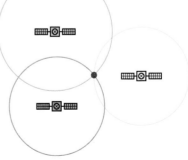

Modern timekeeping

Although the Royal Observatory is the home of Greenwich Mean Time, the public time service at Greenwich came to an end at the start of the Second World War, when it was moved to Abinger in Surrey and later to Herstmonceux in Sussex. In the 20th century, clocks were developed with rates that were steadier and more stable than that of the spinning Earth, which is gradually slowing down.

In 1967, without changing its length, scientists changed the definition of a second from one based on 1/86,400 of an average day to one based on the properties of caesium atoms. In 1972, a new time scale, **Coordinated Universal Time** (UTC), replaced Greenwich Mean Time as the basis of international timekeeping. For most practical purposes they can be regarded as the same thing, because UTC is adjusted so that it always remains within 0.9 seconds of Greenwich Mean Time.

Time today comes from the averaged readings of about 200 atomic clocks located in laboratories around the world. Atomic clocks make use of microwaves similar to those used by mobile phones and microwave ovens. The microwaves are produced inside the clock by an oscillator vibrating 9,192,631,770 times a second. An electronic device then counts the vibrations and converts the count rate into a time on a clock face. Every time 9,192,631,770 vibrations are counted, the time shown by the digital display increases by one second.

DID YOU KNOW?
Days are about two hours longer than they were 400 million years ago. This is because the gravitational pull of the Moon exerts a force on the Earth, which means each spin is ever so slightly slower than the last. By the year 2100, a day on Earth will be about 2 milliseconds longer.

▲ The laser satellite tracking station at Herstmonceux is one of a number of satellite tracking stations located around the world. By using a laser beam to measure the distance to orbiting satellites in known and stable orbits, variations in the Earth's spin rate can be measured, along with tidal movements of the land and continental drift.

The International Earth Rotation Service in Paris coordinates information from satellite tracking stations and other sources about variations in the Earth's spin rate. Adjustments of a second at a time are made to keep the UTC time scale in step with the gradually slowing Earth. In the first 20 years of UTC, a total of 22 days had an extra second added to them. These extra seconds are called leap seconds.

▲ This atomic clock, dating from the 1990s, was designed to keep time to the equivalent of 1 second in 1,600,000 years.

DID YOU KNOW?
Clocks are currently being developed with a theoretical accuracy equivalent to losing or gaining no more than one second in the lifetime of the universe.

Still on the line...

Although our lives have changed significantly with new technologies in navigation and timekeeping, the Royal Observatory is a powerful reminder of our enduring relationship with the Sun, Moon and stars. The historic Prime Meridian is just one of many meridians that were established here to help improve navigation and save lives at sea. Using a combination of clocks and telescopes, generations of Astronomers Royal and their assistants spent many cold nights carefully observing the stars to translate the Earth's daily rotation and motion through space as a measure of time and location. They collaborated with the best clockmakers and craftsmen to design the most sophisticated instruments possible, creating astronomical books and time standards that navigators relied on. After two centuries of work, the adoption of the Airy Meridian as the world's Prime Meridian in 1884 confirmed the Observatory's global significance that literally put Greenwich on our maps, globes and clocks. Even today, we are still 'on the line'.

About the Royal Observatory, Greenwich

One of the most important sites in the world, the Royal Observatory is the historic home of British astronomy, Greenwich Mean Time and the Prime Meridian. It was founded by Charles II in 1675 and now welcomes visitors around the globe to stand at the centre of world time.

Become a Member

Become a Member of Royal Museums Greenwich and enjoy free entry to the Royal Observatory, *Cutty Sark*, Planetarium shows and special exhibitions. In addition, receive invitations to exclusive Members' events, benefit from discounts on other events, and enjoy a 10% discount in our cafes and shops.

The valuable support of our Members, Patrons, donors and sponsors allows us to continue our important work, through exhibitions, our website, loans, conferences, publications, learning programmes and community initiatives. For more information, visit rmg.co.uk/membership

Picture Library

The Royal Museums Greenwich Picture Library contains over 45,000 images and photographs depicting notable ships, empire, trade and exploration, navigation, astronomy and time. All the images within this book are available to license, please contact pictures@rmg.co.uk for more information.